laserpronet.com
Empowering the Laser Workforce

What we do at LaserPronet

- Screening Exams
- Professional Development Courses
- Professional Growth Plans
- Certifications

Copyright © 2018

Sukuta Technologies, LLC

All rights reserved

Table of Contents

1. Laser Technician Series   4
2. Solid State Laser Optical Pumps and their Common Problems Notes   7
3. Solid State Laser Optical Pumps Self-Test   35

1. Laser Technician Series

Laser Technician I Job Description: Level I Technicians perform final tests on lasers and laser systems to ensure that they fully comply with customer specifications. All tests are completely documented for both internal and external use.

Environment: Test Laboratory.

Requirements: Must understand laser and optics principles and, laser beam performance specifications. Experience testing/evaluating laser beams and working with photonics test equipment is a must. Also, ability to follow prescribed written procedures, directions and strict adherence to best laser lab and manufacturing practices are required.

Minimum Educational Requirements: Candidate must hold a certificate/degree in Laser/Electro-Optics Technology or related discipline

Laser Technician II Job Description: Level II Technicians assemble and troubleshoot common optical problems in laser systems. They perform final tests on the systems to ensure that they fully comply with customer specifications and all tests are completely documented for both internal and external use.

Environment: Manufacturing and Test Lab

Requirements: Must understand Wave and Geometrical Optics, Gaussian Beam Propagation, Nonlinear Optics and how acousto- and electro-optics modulators, optical components and accessories work. Experience aligning laser systems, troubleshooting laser beam aberrations and working with photonics test equipment is a must. Also, ability to follow prescribed written procedures, directions and strict adherence to best laser lab and manufacturing practices are required.

Minimum Educational Requirements: Candidate must hold a certificate/degree in Laser/Electro-Optics Technology or related discipline.

Laser Technician III Job Description: Level III Technicians assemble, align, burn-in, test, and tune/troubleshoot laser heads until they meet all performance specifications.
Environment: Manufacturing
Requirements: Must understand the fundamentals of solid state laser technology, accessories, and support systems. Experience aligning laser systems and cavities/resonators and using photonics test equipment is a must. Also, ability to follow prescribed written procedures, directions and strict adherence to best laser lab and manufacturing practices are required.
Minimum Educational Requirements: Candidate must hold a certificate/degree in Laser/Electro-Optics Technology or related discipline

Level 3 Volumes - Solid State Laser Basics and their Common Problems
Volume 1: Solid State Laser Optical Pumps
Volume 2: Amplifying Crystals
Volume 3: Laser Resonator Longitudinal Modes
Volume 4: Laser Resonator Transverse Modes
Volume 5: Laser Oscillator Efficiencies

Technician IV Job Description: Level IV Technicians support Research and Development (R&D) and, customers. Customer Support/Technical Service Technicians work on deployed lasers and laser systems in support of existing customers. R & D/Engineering Technicians support scientists and engineers improve existing, and also create the next generation laser technologies. Level IV Technicians work under limited supervision.
Environment: Research/Engineering Lab and Field
Requirements: Must have a thorough understanding of solid state laser technologies and support systems, experimentation and research protocols. Experience collecting and analyzing data, troubleshooting/problem-solving, using MS Office to generate test

reports and writing technical reports is required. Customer Support/Service technicians may have to travel to customer sites. **Minimum Educational Requirements:** Candidate must hold a certificate/degree in Laser/Electro-Optics Technology or related discipline.

1. Solid State Laser Optical Pumps and their Common Problems
Laser Technician Level III Volume 1

Table of Contents

A. Laser Power Supplies Overview	8
B. Laser Pumping Schemes	11
B.1 Optical Pumping	
B.2. Electrical Pumping	
C. Solid State Laser Optical Laser Pumps	13
C.1 Laser Lamp Pumps	
C.1.1 Elliptical Pump Lamp Cavities	
C.1.2 Flash lamp Lifetime	
C.2 Solid State Laser Diode Laser Pumps	17
C.2.1 Laser Diode Mechanisms and Output	
C.2.1 Laser Diode Issues	

A. Laser Power Supplies Overview

- A power supply converts AC to DC electricity before supplying it to the laser pump
- Components of power supply include a transformer and bridge rectifier

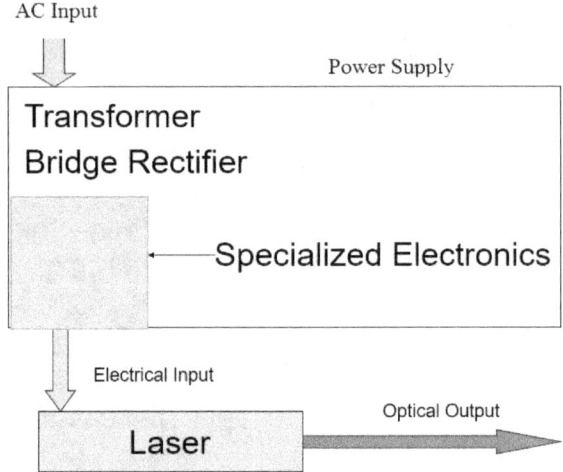

- A transformer transfers primary electricity from wall electrical outlets to either a higher (step-up) or lower (step-down) secondary voltage depending on a laser's voltage requirements.
- Ideally power is conserved in a transformer
    - $\text{Power}_{primary} = \text{Power}_{secondary}$
- Power=Voltage x Current
    - So if voltage goes up in the secondary circuit, the current will go down.

- Transformers do not work with DC electricity, so the up/down conversion must happen before the bridge rectifier

- The bridge rectifier converts AC to DC electricity

- A "clever" combination of rectifiers called a Bridge Rectifiers.

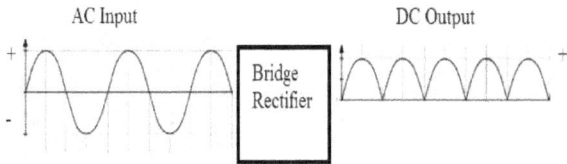

- Laser pumps use DC.
- The more linear the output DC is the less AC noise the laser signal/beam will have

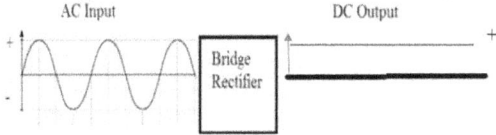

- Electrical signals in a laser beam can be recovered by a sensor and analyzed for noise
- The beam signal will be converted back to electrical and decomposed into AC and DC components by the analyzer.
- The "good" laser signal is the DC component of the extracted signal
- Any AC signals recovered in a laser beam are treated as noise e.g. RMS and Peak-to-peak Noise.
- It is therefore prudent to check/verify the AC noise of a power supply before using it on a laser.
  - Please note that there could be other sources of electrical noise in a laser beam.

A.1 Power Supply Conversion Efficiency
- No power supply is 100% energy conversion efficient because
    - no real transformer ever recovers all the energy/power from the primary
    - the less AC noise (rms, peak-to-peak) the power supply has the less energy efficient a bridge rectifier/power supply will be.

These and other energy conversions and signal "clean-ups" within the power supply adversely impact the wall-plug efficiency of a power supply and subsequently of the laser or laser system.

Personal Notes

B. Laser Pumping Schemes

- In thermal equilibrium electrons always occupy the lowest energy levels (Ground State) in all materials, including laser active media.
- A laser active medium will not lase in thermal equilibrium

- A power supply provides electrical energy to a laser pump
- A laser pump transfers that energy to the laser active medium within the resonator
- The transfer of excitation energy to the active medium is called pumping

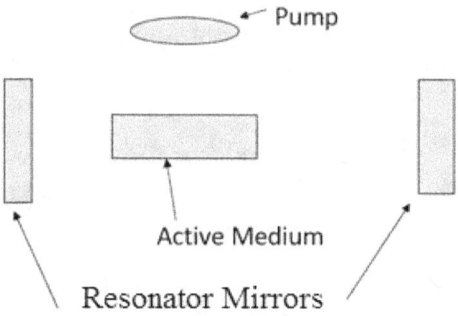

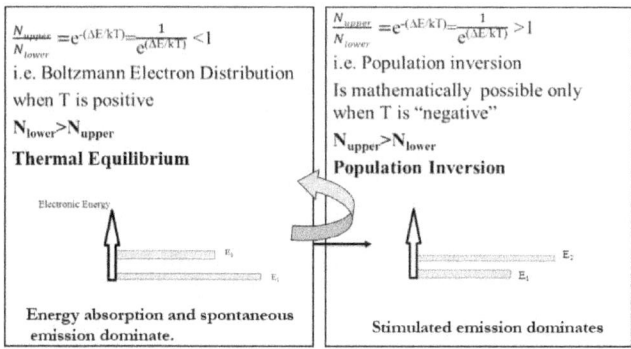

- After pumping a laser active medium more electrons should occupy a higher energy level (E_2), compared to an energy level below it (E_1), and this is called population inversion.

- Two common methods for laser pumping involve the injection of either optical or electrical energy in a laser active medium.

    ### B.1 Optical Pumping

    - Involves the exposure of the active medium to electromagnetic radiation to create population inversion and maintain population inversion
    - The active medium must be transparent, or at least to some extent.

    ### B.2. Electrical Pumping

    - Involves the passing of an electrical current through the active medium to create and maintain population inversion
    - The active medium must be electrically conductive

C. Solid State Laser Optical Pumps

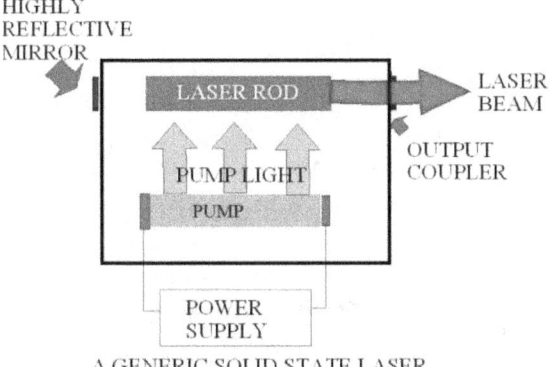

A GENERIC SOLID STATE LASER

- Optical pumps are used to excite both solid state and semiconductor laser active media (OPSL)
- Flash lamps, cw arc lamps, laser diodes are used as laser optical pumps

- When a **d**iode laser **p**umps a **s**olid-**s**tate the laser the laser is called a **DPSS** laser
- When a lamp pumps a solid-state material, the laser is called lamp-pumped solid-state laser.

a. Laser Lamp Pumps

- The solid-state laser pump lamps are primarily flash and arc lamps

- A flash lamp consists of a sealed glass tube with a gas mixture inside and then electrodes to transmit DC electricity through it.
- Typical gases inside a flash lamp sealed tube include Krypton and Xenon
- The electrodes are called anode (+) and cathode (-)

- Components of a flash lamp power supply would also include a Pulse Forming Network

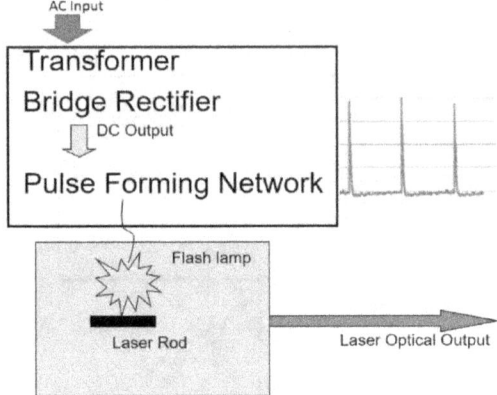

- An ionizing high voltage source energizes a flash lamp gas mixture, thereafter the ionized gas is electrically excited to fluoresce/glow
    - The process is identical to the excitation of ion gas lasers
- The gas ionization and excitation process also produces intense heat within the sealed tube
- Cw arc lamps operate on the same principle as "lightning".

C.1.1 Elliptical Pump Lamp Cavities
- Laser lamp pumps are housed in reflective elliptical cavities

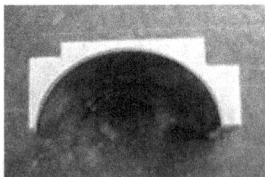

Figure. End-view of one half of an elliptical cavity

- The laser/active medium and lamp are each placed at one of the foci of the elliptical cavity
- Lamps allow for mostly indirect introduction of pump light to the active medium

- Walls of elliptical laser cavity are shiny so that lamp light from one foci can be directed to the other foci within the active medium upon reflection.

C.1.2 Flash lamp Lifetime

- Flash lamp lifetime can be increased and pump efficiency improved if the lamp is pre-ionized
- An auxiliary power supply is needed for simmer mode laser operation
- In the simmer mode, a low current discharge is maintained in the lamp by a simmer power supply

- Liquid cooling of lamps makes it possible to operate them at maximum inner-tube wall thermal loadings
- A lamp pump is replaced whenever it ruptures and/or reaches the end of its useful lifetime
- A lamp will rupture if used past its useful lifetime and/or if its poorly handled
- If a lamp ruptures the cavity will have be to be cleaned and realigned

Personal Notes

C.2 Solid State Laser Diode Laser Pumps

- Semiconductor electronic energy band gaps are not as narrow as conductors but also not as large as insulators - they are intermediate

- They are very poor conductors at low temperatures, so the electrons will have very low kinetic energy.

- They conduct electricity at higher temperatures (kT) as electrons get thermally excited into their Conduction Bands

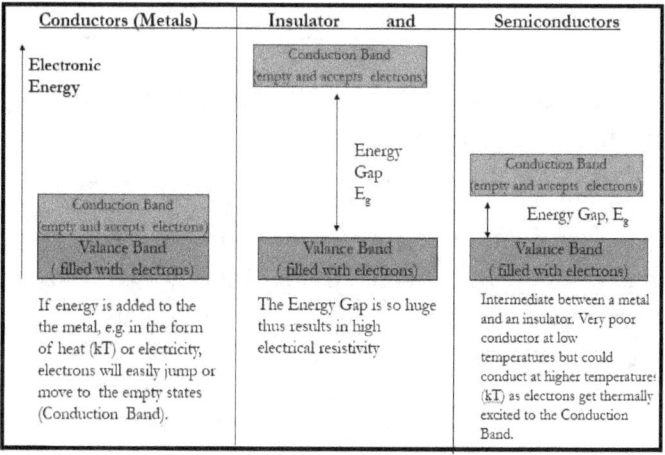

Partial List of Semiconductors

	Crystal	Chemical Symbol
1	Silicon	Si
2	Germanium	Ge
3	Indium Phosphate	InP
4	Gallium Phosphate	GaP
5	Gallium Arsenide	GaP

- A semiconductor can be doped with a donor impurity to make available free electrons – i.e. n-type material; here n stands for negative (electrons).

- Also, a semiconductor can be doped with an acceptor impurity to create an electron deficiency – i.e. p-type material, here p stands for positive charge also known as a "hole"

- **Example of n-type material: Silicon Arsenic Doping**
    - If silicon is doped with an arsenic atom, As $[Ar]3d^{10}4s^24p^3$ with 5 valence electrons, an extra electron would be created
    - 5(Arsenic) +4(Silicon)=9 electrons
    - 8 electrons are needed to have a stable configuration and this covalent bond has one extra electron
    - The extra electron is let "loose" into the semiconductor crystal thus create a n-type material

- **Example of p-type material: Silicon Boron Doping**
  - When silicon is doped with a Boron atom, Boron ($1s^2 2s^2 2p^1$) with 3 valence electrons, one electron would be missing to create a closed-shell stable configuration.
  - 3(Boron) +4(Silicon)=7 electrons
  - 8 electrons are needed to have a stable configuration and this covalent bond is missing 1
  - A "hole" or positive charge (p-type) is therefore created the crystalline structure.

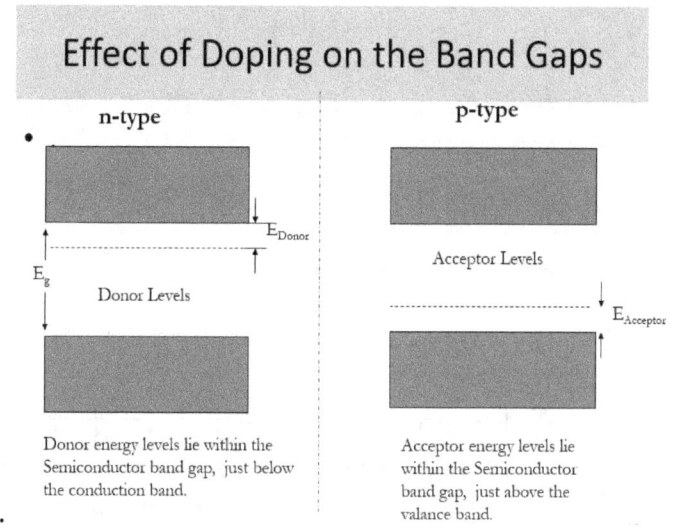

C.2.1 Laser Diode Mechanisms and Output

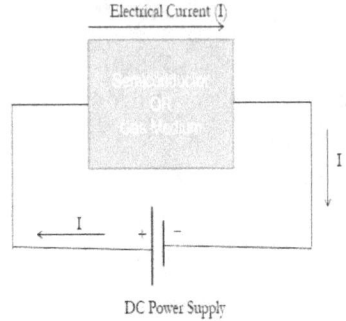

Passing current in through the active medium transfers energy to the electrons

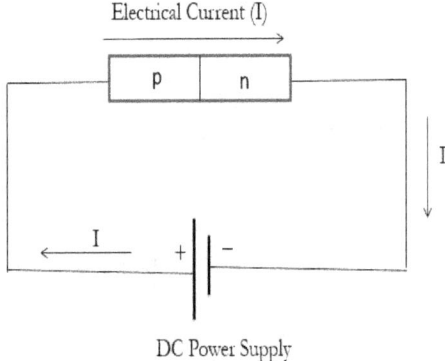

Passing current through the active medium transfers energy to the electrons

- Electrons absorb electrical energy to ascend from the valence to the conduction band.
  - The electrons are naturally bound in the valence band and they ascend/rise to the conduction band after absorption of electrical energy.

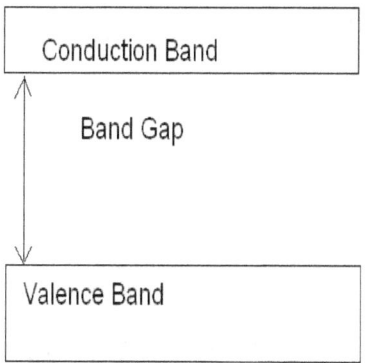

- To produce light electrons traveling in the n-region fall into holes in the p regions and photon emission occurs.

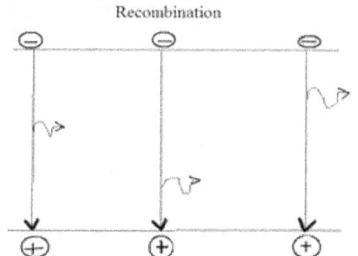

An electron releases energy to recombine with a hole

- The distance between the two bands is called the energy band gap (E_g)

    - $E_{gap} = \dfrac{hc}{\lambda}$
        Where h is Plank's constant
        - c the speed of light, and
        - λ the output wavelength

- Light is produced spontaneously and this is how an Light Emitting Diode (LED) works

P-n Junction LED

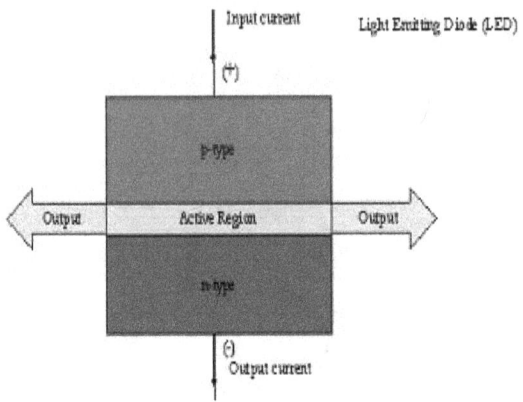

LED Emission is Random

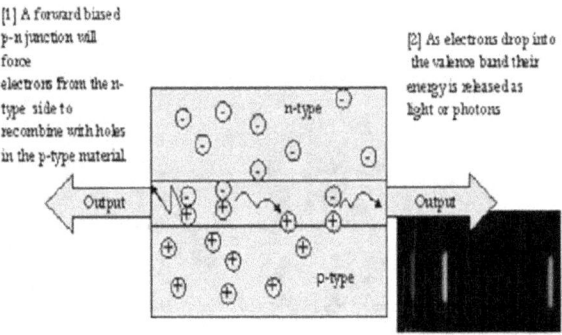

LED output has multiple wavelengths or is multiline

- After inserting HR and OC the device functions as a laser diode/diode laser whereby
  - The passing electric current would create population inversion in the conduction band

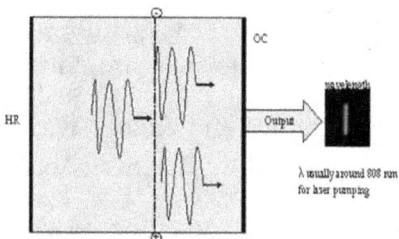

Semiconductor Lasers are also known as laser diodes or diode lasers

- Semiconductor laser include Doped GaAs , InGaAsP , AlGaInP GaN, etc..
- Semiconductor lasers are also known as laser diodes/diode lasers

Laser Diode Performance and Attributes

- The intensity/power of the produced laser light is proportional to the pump drive current
- Laser diode light energy/wavelength output is directly to proportional to the energy band gap
  - The NIR wavelength output range is within the ND-doped, and other solid-state, active media absorption bands
  - Therefore, the laser diodes are used as pumps for solid-state laser media where they allow for
    - direct injection of pump energy onto the active medium
    - matching of their spectral output to media absorption bands
      - The most common laser diode pump wavelength for pumping Nd:YAG and ND: YVO_4 lasers is around 808nm at room temperatures
- Laser diodes have high electrical-to-optical efficiency range averaging around 60%
  - This is much higher than the efficiency of lamps
- Usually a single laser diode does not sufficient energy/power to drive most applications, so they are packaged in "gangs" called Laser Diode
  - Arrays
  - Stacks
  - Bars

- Laser diodes
    - generate very low heat
    - have a small footprint (~mm³) and volume, and weigh much less compared to lamps

C.2.2 Laser Diode Issues
C.2.2.1 Output Wavelength Fluctuations
C.2.2.2 Sensitivity to electro-static discharge (ESD)
C.2.2.3 Asymmetric Beam Divergence

C.2.2.1 Output Wavelength Fluctuations
- A typical laser diode will produce heat while in operation.
- Sources of heat in laser diodes include
    - p-n junction current
    - ambient/housing set temperature
- Moreover, temperature affects the laser diode energy gap as demonstrated in the experimental results shown below

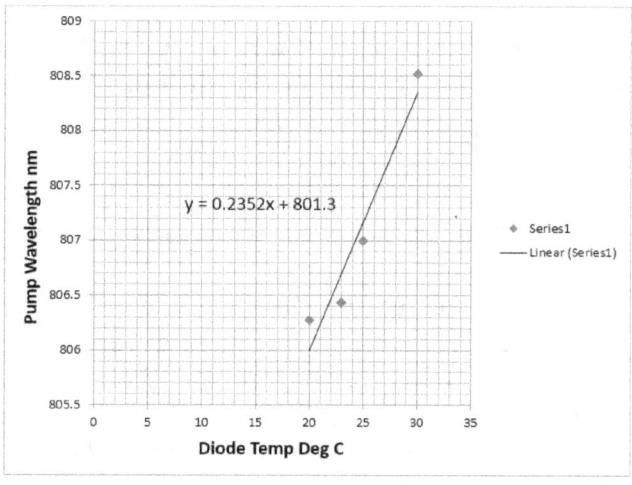

- The graph demonstrates that temperature affects the energy gap and thus subsequently affect or alter the output wavelength so we can now recast the energy bandgap equation as
  - $E_{gap}(T) = \frac{hc}{\lambda} = \frac{1.24\,um}{\lambda}$
- This means that laser diode wavelength must be specified at a specific temperature
  - For example, λ=808 nm @ 28.0 °C
- In addition, laser diode temperature must "locked in" at that specific temperature to avoid fluctuations if a constant output wavelength is desired.
- Laser diode life decreases with increasing temperature.
- To keep the temperature of a laser diode constant the housing must either be cooled/warmed with the aid of a heat sink
  - Unfortunately, when a heat sink lowers temperature this could induce condensation within the housing which may degrade the laser diode's performance.
  - On the other hand, laser diode performance will decrease with increased operating temperatures
  - It is of course best to get a laser diode package that outputs the wavelength you want at the application ambient temperature.

Thermal Management for Wavelength Control

- The heat generated by laser diodes must be monitored and controlled for optimal operations
- Thermistors and RTDs can be used as temperature sensors in lasers.
  - RTD is the acronym for Resistance Temperature Detectors
- A TEC can maintain optimal temperature
  - TEC acronym stands for Thermal Electric Cooling. It is also known as at the Peltier Effect.
  - A thin film of thermal grease is generally applied between the TEC and surface to be cooled/heated

Personal Notes

Laser Diode Failure

- Infant Mortality Failure- are caused by defects in the manufacturing process which may also be caused by poor screening of raw materials
- Environmental-induced Failure - include ESD, electrical and thermal instabilities.
    - Appropriate heat extraction systems and drive currents must always be applied
    - These failures impose a constant threat throughout the lifetime of the laser diode/semiconductor device
- Care must be taken to ensure that the laser diodes are operated in a static-free (ESD-free) environment
    - Grounded workbench/surface
    - Grounded wrist strap on the user. User must always be grounded when working with laser diodes or any micro-circuits
        - ESD wrist straps should be worn whenever handling electronic/electronic micro-devices and components
        - They provide an alternative low-resistance path for electrons to flow *in lieu* of a micro-circuit such as those found in laser diodes
    - Anti-static floors

- Technician/user must also wear ESD reducing attire such as shoes and gown etc.
- Use of topical applications also help reduce ESD
- Keep the laser diode in static-free containers when not in use
- Wear and tear, usage etc.. are a consequence of aging/time and their probability increases with time.

Reliability Tests are needed for burn-in and lifetime tests.

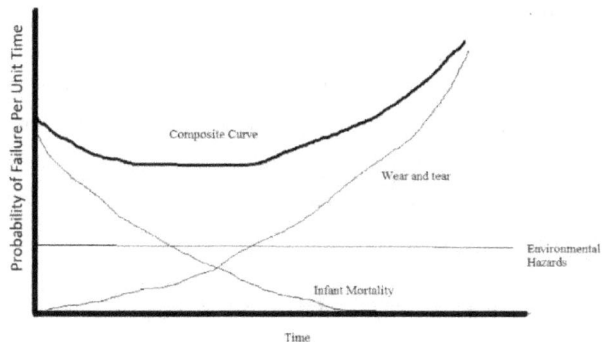

C.2.2.2 Sensitivity to electro-static discharge (ESD)

Tribo-electric Series

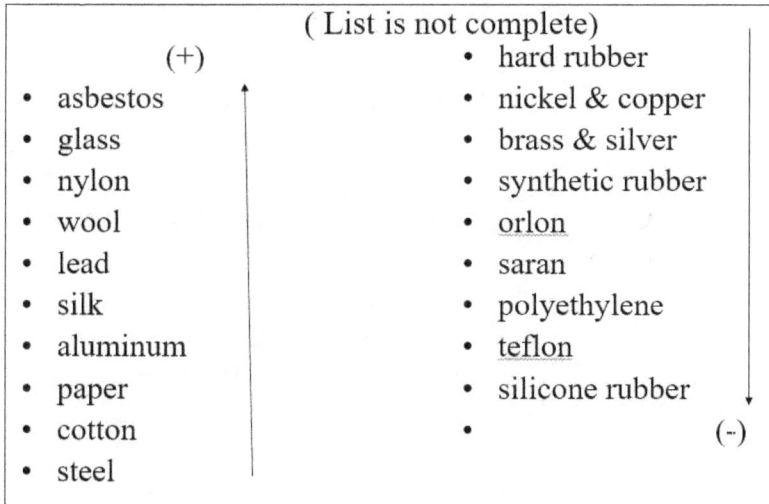

(+)	(List is not complete)
• asbestos	• hard rubber
• glass	• nickel & copper
• nylon	• brass & silver
• wool	• synthetic rubber
• lead	• orlon
• silk	• saran
• aluminum	• polyethylene
• paper	• teflon
• cotton	• silicone rubber
• steel	• (-)

- When rubbed together the material that charges positive will be the one that is closer to the positive end of the series

- The material closer to the negative end will charge negatively.
- Materials generally strive to be electrically neutral by
  - either accepting electrons, if its positively charged already OR
  - discharging electrons if its negatively charged already

Microcircuits and ESD

- The human body accepts and discharges currents/electrons to stay electrically neutral.
- The movement of charge to, or from, materials to achieve neutrality is called Electrostatic Discharge (ESD)
- The magnitude of ESD currents is sufficient to damage laser diodes or any micro-circuits.
  - Electronic/electronic micro-devices and components include PCBs and laser diodes
  - PCBs stands for Printed Circuit Boards
  - An electronically "unbalanced" body will cause electrons/current to pass through a microcircuit if in contact with a third body/object
  - Micro-circuits can be damaged if current flows through them
  - It is therefore important that precautions are taken while handling laser diodes to avoid ESD damage.

C.2.2.3 Asymmetric Beam Divergence

- Edge-emitting laser diode output beams are always elliptical and uncollimated
- Semiconductor laser diodes have highly divergent and elliptical beams due to uneven diffraction in two dimensions

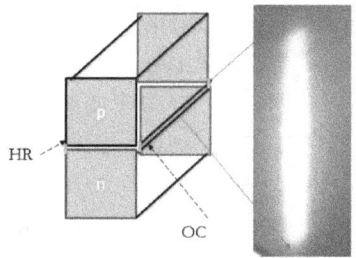

- The pn junction of an edge-emitting laser diode creates divergent asymmetric/uneven beams.
    - Divergent beams do not have precise focal points
    - On the other hand, asymmetrical beams will produce two focal points when focused i.e. astigmatism
        - Vertical component ($\theta_y = k\lambda/d_y$) is more divergent than the
        - Horizontal components ($\theta_x = k\lambda/d_x$) is less divergent
            - since $d_y < d_x$

  - Astigmatism is due to the different x and y convergent points when focused
  - Beam Divergence- two divergences in orthogonal planes requires anamorphic optics i.e. optics with different focusing properties in the two planes

Beam Collimation

- A positive lens is placed one focal length from exit port of the edge-emitting laser diode
- This would collimate the asymmetric beam only.

Beam Circularization Optics

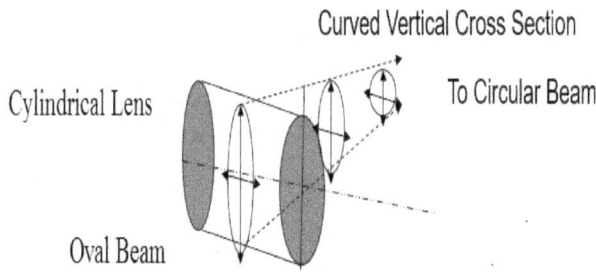

- Anamorphic optics are generally used to circularize asymmetric beams
    - They can either expand the shorter axis or shrink/contract the longer axis of the laser beam.
    - Anamorphic optics include
        - cylindrical lens
        - or a pair of prisms,

- Once the asymmetry has been corrected the focused beam will not exhibit astigmatism when focused
- A collimated pump beam focused into the active medium/laser rod is desired

End-pumping DPSS Lasers

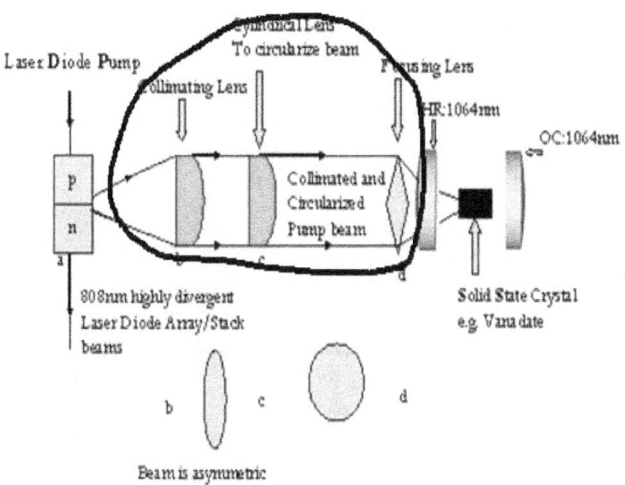

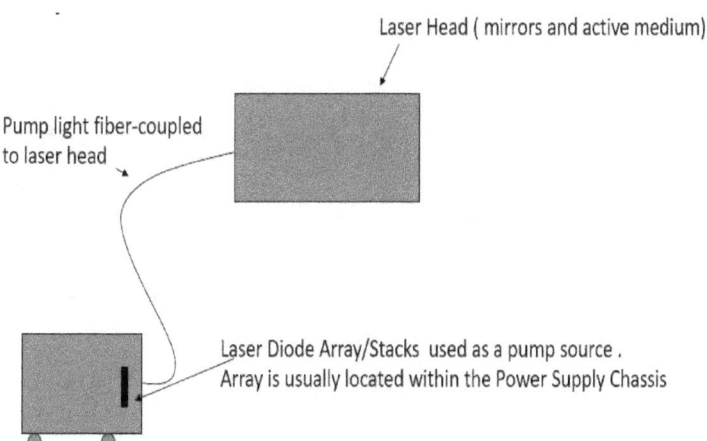

Vertical Cavity Surface Emitting Lasers (VCSEL)
- VECSEL *(vixel)* stands for Vertical Cavity Surface Emitting Laser
- VCSEL produce circular beams but they have not been adopted in most DPSS lasers.

Personal Notes

2. Solid State Laser Optical Pumps their Common Problems Self-Test

When done go to laserpronet.com and test your knowledge

Table of Contents   Page

    A. Basic Laser Scheme   36
    B. Laser Pumps' Power Supplies   40
    C. Solid State Laser Optical Pumps   42
    D. Solid State Laser Lamp Pumps   45
    E. Solid State Laser Diode Laser Pumps   49

A. Basic Laser Scheme

1. A laser consists of a(n) optical
    a. resonant cavity
    b. gain/amplifying /active medium
    c. all the above
    d. none of the above

2. A_____must support oscillations necessary for the stimulated emission/de-excitation of the inverted electron population.
    a. gain/amplifying/active medium
    b. Pump
    c. laser cavity/resonator
    d. None of the above

3. A laser pump_____.
    a. regulates the temperature of the laser
    b. pumps coolant to a laser cavity
    c. allows some of the laser light to leave the cavity to produce the laser's output beam.
    d. transfers energy to the laser active medium
    e. None of the above

4. A laser resonant cavity consists of_____mirrors
    a. one
    b. two
    c. three
    d. any of the above
    e. None of the above

5. A laser mirror surface must be _____ to the axis is of the laser resonator.
   a. parallel
   b. perpendicular
   e. None of the above

6. A laser_____must create and maintain population inversion in the active medium for specific output wavelength
   a. mirrors(s)
   b. gain/active medium
   c. pump
   d. All the above
   e. None of the above

7. A(n)_____absorbs external energy to create an electron population inversion.
   a. gain/active/amplifying medium
   b. pump
   c. laser cavity/resonator
   d. All the above
   e. None of the above

8. Electrical pumping in gas lasers involve the discharge of high-velocity_____between electrodes.
   a. electrons
   b. protons
   c. neutrons
   d. all the above
   e. None of the above

9. Electrical pumping of gaseous laser/active media involves_____electronic excitations.
   a. collisional
   b. gravitational
   c. a and b
   d. None of the above

10. Laser gaseous active media could be _____
    a. atomic
    b. molecular
    c. a and b
    d. None of the above

11. The goal of electrical pumping of gaseous active media could include _____
    a. ionization
    b. nuclear excitation
    c. electronic excitation
    d. a and/or c
    e. Bone of the above

12. An excited laser active medium could undergo electronic _____.
    a. population inversion
    b. thermal equilibrium
    c. a and b
    d. None of the above

13. The transfer of energy from an an electrical current to the laser/active medium to create population inversion happens in _____ active media
    a. solid-state
    b. gaseous
    c. semiconductor
    d. a and b
    e. None of the above

14. In_____lasers, electrical pumping is accomplished by passing a current through the active medium
    a. diode
    b. semiconductor
    c. gas
    d. all the above
    e. None of the above

15. _____ can be electrically or optically excited.
    a. Semiconductor lasers
    b. Laser diodes
    c. Diode lasers
    d. All the above
    e. None of the above

16. A(n) _____ provides electrical energy to a laser pump.
    a. q-switch
    b. power supply
    c. HR
    d. OC
    e. None of the above

17. Most pump energy that enters a laser emerges as _____.
    a. laser radiation
    b. waste heat
    c. magnetic energy
    d. all the above
    e. None of the above

18. Laser pump energy imparted to the active media is converted _____ into laser radiation.
    a. 100%
    b. Less than 100%
    c. 50%
    d. All the above
    e. None of the above

B. Laser Pumps' Power Supplies

19. A power supply_____.
    a. eliminates electrical noise from a laser beam.
    b. creates magnetic energy for a laser
    c. regulates the temperature of the laser
    d. converts AC from the wall into DC electricity before supplying it to the laser pump
    e. None of the above

20. The components of a flash lamp power supply include a _____.
    a. Transformer
    b. Bridge Rectifier
    c. Pulse Forming Network
    d. All the above
    e. None of the above

21. In a laser power supply, a rectifier circuit converts _____ electricity.
    a. DC to AC
    b. AC to DC
    c. DC to digital
    d. None of the above

22. In a laser power supply, a transformer
    a. converts DC to AC electricity
    b. converts AC to DC electricity
    c. steps down/up an AC voltage
    d. steps down/up an DC voltage
    e. None of the above

23. A poorly performing power supply rectifier circuit may result in a laser beam with_____.
    a. huge RMS noise
    b. huge peak-to-peak noise
    c. best M^2
    d. a and b
    e. None of the above

24. A laser power supply with a_____electrical noise will produce laser beams with low noise.
    a. low
    b. high
    c. None of the above

25. Electrical energy can be transformed to_____energy
    a. optical
    b. thermal
    c. kinetic
    d. All the above
    e. None of the above

C. Solid State Laser Optical Laser Pumps

26. Which of the following is/are used as solid-state laser pump(s)?
    a. Flashlamps
    b. Cw arc lamps
    c. Laser diodes
    d. Electrical currents
    e. a, b, and c
    f. None of the above

27. Optical pumps are used to excite_____laser active media
    a. gaseous
    b. solid state
    c. semiconductor
    d. b and c
    e. None of the above

28. Optical pumps facilitate the transfer of electrons in a laser active medium from its ground state to the_____.
    a. Metastable State/Level
    b. Upper Laser Level
    c. Pump Band
    d. a and b
    e. None of the above

29. When a diode laser pumps a solid-state the laser is called a_____laser.
    a. DPSS
    b. LED
    c. Lamp-pumped
    d. Any of the above
    e. None of the above

30. When a lamp pumps a solid-state active medium, the laser is called _____ laser.
    a. DPSS
    b. LED
    c. Lamp-pumped
    d. Any of the above

31. When a laser pumps another laser, it is possible to _____ the pump radiation into the active medium with little or no losses.
    a. directly introduce
    b. focus
    c. a and b
    d. None of the above

32. Laser engineers strive to design optical pump systems that transfer energy to the gain medium with _____ losses.
    a. 50%
    b. minimal
    c. 100%
    d. Any of the above
    e. None of the above

A laser pump's current (x) was increased and the laser output measured (y) as shown below.

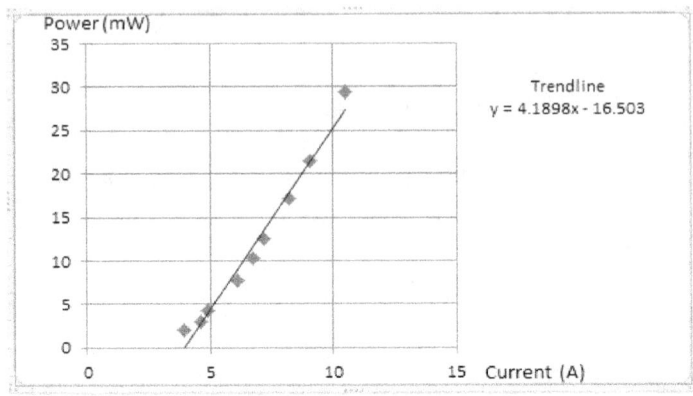

Figure 1. Laser pump effects on output power.

Refer to figure 1 for to answer the two consecutive the questions below

33. the laser threshold at about
   a. 8 A
   b. 8 mA
   c. 4 A
   d. 6 A
   e. None of the above

34. The slope efficiency (mW/A) of the laser is about
   a. 4.2
   b. .4
   c. 16.5
   d. -16.5
   e. None of the above

D. Solid State Laser Lamp Pumps

35. Typical gases inside a flash lamp tube generally include _____.
    a. Krypton
    b. Xenon
    c. a and b
    d. None of the above

36. A flash lamp consists of a _____.
    a. sealed glass tube
    b. gas mixture inside a sealed tube
    c. electrodes
    d. All the above
    e. None of the above

37. A flash lamp gas mixture is energized by a(n) _____.
    a. ionizing high voltage source
    b. non-ionizing high voltage source
    c. magnetic coil
    d. either a or b
    e. None of the above

38. Flash lamp electrodes at the ends of flash-lamps _____ the sealed tube
    a. transmit gas ionizing electrical charge through
    b. excite the gas within
    c. a and b
    d. None of the above.

39. Flash lamp electrodes _____.
    a. are powered by capacitors connected to a Pulse Forming Network
    b. create intense heat in within the sealed tube
    c. a and b
    d. None of the above

40. Laser lamp pumps are housed in reflective _____ cavities
    a. circular
    b. rectangular
    c. oval
    d. elliptical
    e. None of the above

41. Lamps allow for mostly_____introduction of pump light to the active medium
    a. direct
    b. indirect
    c. a and b
    d. None of the above

42. In lamp-pumped laser cavities the active media is installed at_____of the cavity.
    a. one of the foci
    b. both foci
    c. the center
    d. Any of the above
    e. None of the above

43. In lamp-pumped cavities the lamp is installed at the ____of the cavity.
    a. one of the foci
    b. both foci
    c. the center
    d. Any of the above
    e. None of the above

44. Walls of a lamp pumped elliptical laser cavity are shiny so that lamp light can be_____.
    a. refracted to outside the cavity
    b. distributed evenly throughout the cavity
    c. reflected to one of the foci with the laser rod/active medium
    d. None of the above

45. A(n) _____ useful lifetime can be increased if operated in simmer mode.
    a. Flash lamp
    b. Arc lamp
    c. All the above
    d. None of the above

46. An auxiliary power supply is needed for the_____ mode laser operation.
    a. simmer
    b. cw
    c. all the above
    d. None of the above

47. In the_____ mode a low current discharge is maintained in the lamp by a simmer power supply.
    a. simmer
    b. pulsed
    c. cw
    d. any of the above

48. A_____ active medium absorption band is required for efficient arc lamp pumping
    a. broad
    b. narrow
    c. any of the above
    d. none of the above

49. _____ cooling of lamps makes it possible to operate them at maximum inner-tube wall thermal loadings
    a. Air
    b. Liquid
    c. a and b
    d. None of the above

50. The_____ of a lamp output is smaller than that of a diode laser
    a. spectral bandwidth
    b. divergence
    c. All the above
    d. None of the above

51. A lamp pump is replaced whenever it _____
    a. ruptures
    b. reaches the end of its useful lifetime
    c. a and b
    d. None of the above

52. A lamp will rupture if it's _____
    a. used past its useful lifetime
    b. poorly handled
    c. a and b
    d. None of the above

53. If a laser lamp ruptures the laser cavity will have to be_____.
    a. cleaned
    b. realigned
    c. a and b
    d. None of the above

54. Laser lamps could be_____.
    a. arc
    b. discharge
    c. a and b
    d. None of the above

E. Solid State Laser Diode Laser Pumps

55. LED stands for
    a. Light Emitting Diode
    b. Laser Emitting Diode
    c. Light Emitting Device
    d. All the above
    e. None of the above

56. Semiconductor lasers include doped
    a. GaAs
    b. InGaAsP
    c. AlGaInP.
    d. All of the above
    e. None of the above

57. A laser diode does require_____.
    a. HR
    b. OC
    c. population inversion.
    d. All the above
    e. None of the above

58. A laser diode is also known as a_____laser.
    a. diode
    b. semiconductor
    c. LED
    d. a and b
    e. None of the above

59. In semiconductor lasers population inversion can be accomplished through_____pumping
    a. electrical
    b. optical pumping
    c. a and b
    d. None of the above

60. Edge-emitting laser diode beams are naturally_____.
    a. elliptical
    b. circular
    c. collimated
    d. all the above
    e. None of the above

61. _____produce circular beams.
    a. Edge-emitting laser diodes
    b. VCSEL
    c. LEDs
    d. All the above
    e. None of the above

62. The acronym VCSEL stands for
    a. Vertical-Cavity, Semiconductor Emitting Laser
    b. Visual-Cavity, Surface Emitting Laser
    c. Vertical-Cavity, Stimulated Emitting Laser
    d. Vertical-Cavity, Surface Emitting Laser
    e. None of the above

63. VCSEL beams are more_____than those emitted by edge-emitting laser diodes
    a. asymmetric
    b. chromatic
    c. circular
    d. none of the above

64. Semiconductor lasers_____.
    a. are excited by an electric current
    b. can be used to pump solid state lasers
    c. are a component of DPSS lasers
    d. all the above
    e. None of the above

65. Diode lasers are packaged in_____.
    a. bars
    b. arrays
    c. a and b
    d. None of the above

66. Laser pumping solid-state laser active/gain media allows for _____.
    a. direct deposition of pump energy onto the active medium
    b. spectral matching of pump output to gain medium's absorption band(s)
    c. a and b
    d. All the above
    e. None of the above

67. The most common laser diode pump wavelength for pumping Nd:YAG and ND: YVO$_4$ lasers is around _____ at room temperature
    a. 808nm
    b. 1064nm
    c. 532nm
    d. 266nm
    e. none of the above

68. Diode laser _____ depends on temperature.
    a. emission wavelength
    b. threshold current
    c. slope efficiency
    d. All the above
    e. None of the above

69. Diode laser _____ depend(s) on current.
    a. emission wavelength
    b. temperature
    c. slope efficiency
    d. All the above
    e. None of the above

70. When elliptical beams are focused, they produce _____ images
    a. astigmatic
    b. virtual
    c. a and b
    d. None of the above

71. A typical laser diode will produce _____ while in operation.
    a. heat
    b. mechanical vibrations
    c. All the above
    d. None of the above

72. Sources of heat in laser diodes include _____
    a. junction current
    b. ambient/housing set temperature
    c. a and b
    d. None of the above.

73. The heat generated by laser diodes must be _____ for optimal operations.
    a. monitored
    b. controlled
    c. a and b
    d. None of the above

74. _____ can be used as temperature sensors in lasers.
    a. Thermistors
    b. RTDs
    c. a and b
    d. None of the above

75. RTD is the acronym for
    a. Resistance Temperature Detectors.
    b. Radiation Temperature Detectors
    c. All the above
    d. None of the above

76. TEC acronym stands for
    a. Thermal Electric Cooling
    b. Thermistor Electric Cooling
    c. All the above
    d. None of the above

77. A thin film of_____grease is be used to improve the contact between the TEC and surface of the component to be cooled/heated.
    a. thermal contact
    b. current transfer
    c. All the above
    d. None of the above

78. High laser diode temperatures can_____its useful lifetime
    a. enhance
    b. reduce
    c. has no effect
    d. any of the above
    e. None of the above

79. ESD is the acronym for _____
    a. Electronic Discharge
    b. Electrostatic Discharge
    c. Direct Electric
    d. a and b
    e. None of the above

80. _____are not affected by ESD
    a. Laser diodes
    b. Solid state gain media/crystals
    c. a and b
    d. All the above
    e. None of the above

81. ESD wrist straps should be worn whenever handling _____.
    a. Electronic/electronic micro-devices and components
    b. solid state laser crystals
    c. a and b
    d. None of the above

82. Electronic/electronic micro-devices and components include_____.
    a. PCBs
    b. laser diodes
    c. a and b
    d. None of the above

83. PCBs stands for_____.
    a. Printed Circuit Boards
    b. Parallel Circuit Boards
    c. a and b
    d. None of the above

84. Laser diodes have_____ electrical-to-optical/wall-plug efficiency compared to lamps.
    a. low
    b. high
    c. equal
    d. variable

www.ingramcontent.com/pod-product-compliance
Lightning Source LLC
Chambersburg PA
CBHW050024230526
45470CB00003B/1122